# MINES AND MINING

IN

# PENNSYLVANIA

BY

**FRANK HALL**

Chief, Department of Mines

HARRISBURG, PA.:
WILLIAM STANLEY RAY, PRINTER
1904

anon

Annex
TN
24
P4
H17

# MINES AND MINING IN PENNSYLVANIA

By FRANK HALL, *Assistant Chief, Department of Mines*

To speak of the mining industry of Pennsylvania is to speak of her development and progress in all lines of commercial activity  Coal is the foundation on which, with magic-like rapidity and unexampled success, has been reared a superstructure of industries so varied in character and tremendous in scope that they justly claim the admiration of the world

The history of the development of coal and its application to the needs of mankind is replete with interest  At first rejected, misunderstood and unappreciated, this great creative force has in a comparatively brief space of time completely revolutionized the industrial systems of the world, and is to-day the power on which civilization largely depends for the maintenance and continuance of commercial prosperity  England, the long-time undisputed industrial mistress of land and sea, at-tained her glory through the timely and intelligent use of her coals  Germany, the growing giant of recent years, owes to coal her rapidly acquired distinction as a manufacturing nation  And America, the youngest of the trio and the last to enter the race, can attribute to this dynamic power her acknowledged leadership in the great battle for supremacy  Truly it has been said that "coal is the great factor in everything we do," and equally true is it that the terms "iron age," "steel age," "age of steam," may all be translated the "age of coal "  To contemplate the results that would inevitably follow the elimination of this beneficent source of power from modern society, is to draw an appalling picture of deprivation, stagnation and ruin  The machinery of our countless factories would be stopped, the great ocean liner would become the plaything of the waves, and the railway train that bears to market the produce of our land would be left inert and helpless

In view of the inestimable value of coal and its transcendent importance as a means of perpetuating and increasing the mechanical and manufacturing achievements of the present, it is gratifying to note the prominence of Pennsylvania as a producer and user of this universally applied fuel  Pennsylvania is the greatest depository of coal on the North American Continent and has more capital invested in mineral operations than any other state in the Union  This brings into exercise a corresponding amount of productive industry, and it has been practically demonstrated to the commercial and manufacturing interests that our great Commonwealth has in her possession resources of concentrated wealth that place her in a position of enviable superiority, and which, with increasing development and larger utilization, bespeak for her an ultimate elevation that bewilders the imagination

The coal fields of Pennsylvania belong to the great Appalachian coal measures, the first in importance in North America, covering 70,000 square miles and extending from Canada to Alabama  The Pennsylvania coal fields cover an area of about 15,500 square miles and are divided into two great regions, the anthracite in the eastern part of the State, containing about 500 square miles, and the bituminous in the western part, containing about 15,000 square miles  The coals from the two sections differ greatly in appearance, composition and adaptability  The anthracite is distinguished by its compactness, high specific gravity, semi-metallic luster, a preponderance of carbon, and the absence of sulphur and water as constituent parts  It burns with a very small amount of flame, produces intense heat and no smoke  It is largely used in the smelting of iron in air or blast furnaces, where a high temperature is required, and it is also the ideal fuel for domestic purposes  By reason of the great percentage of carbon, it is superior to any other mineral fuel

The bituminous coal is soft and dull in appearance, contains much less carbon, but is richer in hydrogen  This is the most important class of coals, because of the great deposits to be found in almost every country and the manifold uses to which it is applied

No sharply defined line of demarcation can be drawn between the anthracite and bituminous coal fields, as the one series merges by imperceptible degrees into the other  This gradation is observable in the coal deposits themselves, anthracite and bituminous coal being frequently found not far removed in different parts of the same seam, and the gradual transformation from a flaming coal to a compact, lustrous, non-flaming kind being very easily traceable  The variations in composition are attended with corresponding differences in qualities  The first practical use made of anthracite coal, of which we have record, was in 1768, when Obadiah Gore, a blacksmith in the Wyoming Valley, successfully burned it in his forge  From the time of his experiment the coal trade grew, imperceptibly at first, but soon with rapid strides  The principal obstacles to its immediate adoption were the difficulty of ignition and the cheapness and abundance of wood

In 1812, Colonel George Shoemaker, of Pottsville, in his desire to introduce anthracite coal to the people of Philadelphia, loaded several wagons at his mine at Centerville in the Schuylkill region, and proceeded to the city to find a market  The people of Philadelphia had not been favorably impressed with the quality of anthracite coal, and the frequent attempts to impose "rocks" upon them for coal had aroused their indignation  Colonel Shoemaker was therefore heartily denounced as a scoundrel, and in his attempt to introduce a fuel that has since made Philadelphia one of the most prosperous cities in the world, he lost both time and money  The persons to whom he had given coal obtained a writ from the authorities of the city for his arrest as an impostor and swindler, and in order to escape persecution he was forced to take a rapid flight and make a wide circuit around the Quaker City on his way home

Mr White, of the firm of White & Hazzzard, of the Fairmount Nail and Iron Works, was anxious to succeed in burning this coal and he and some of his men made a persistent attempt to burn it in one of the furnaces They spent a whole morning and tried every expedient that skill and experience could suggest They raked the coal, they poked it and stirred it up and blew upon the surface through open doors Their persevering efforts, however, were of no avail Colonel Shoemaker's "rocks" would not burn Dinner time having arrived, the men shut the furnace doors and left in disgust Returning from dinner they were astonished by the phenomenon they beheld The doors of the furnace were red hot with a heat never before experienced The fire had simply been let alone, and "Let it alone" became the motto for the use of this coal thereafter The result of this success was mentioned in the papers and anthracite coal soon obtained a reputation and found friends and advocates in Philadelphia

The anthracite coal fields are divided into three great divisions the Northern or Wyoming, containing about 200 square miles, situated in Lackawanna, Luzerne, and a small part of Susquehanna counties, the Middle field, divided into the Eastern or Upper Lehigh, and the Western or Schuylkill region, containing about 130 square miles in Luzerne county, with small sections in Northumberland, Carbon, Schuylkill and Columbia counties, the Southern field, containing about 140 square miles in Carbon, Schuylkill and Dauphin counties

The first discovery of bituminous coal in Pennsylvania was made early in the Eighteenth century, but the trade in it as a commodity of commerce dates from 1784, when the Penns, who still retained their proprietary interests in the State, including the manor of Pittsburg, surveyed the town of Pittsburg and at the same time sold the privilege of mining coal in the "great seam" opposite the town, at the rate of 30 pounds for each mining lot, extending back to the center of the hill The coal mined at that time and up to 1850 was floated to market down the Ohio River, during the spring and fall freshets, in large flat-bottom boats holding about 15,000 bushels each In 1850 this primitive method was superseded by the introduction of steam tow-boats

The first shipment of bituminous coal to eastern market was made from Clearfield county in 1804 In that year an ark load was sent down the Susquehanna River to Columbia, a distance of 260 miles The new fuel was a great surprise to the people of that section In 1828 the first cargo reached Philadelphia from the same county The means of transportation, however, were too imperfect to permit building up a coal trade between the Alleghenies and the seaboard, and not until some years after the completion of the internal improvements of Pennsylvania was the trade put upon a permanent basis

In 1826, under authority conferred by an Act of the General Assembly, the great project of constructing the Pennsylvania Canal was undertaken   The purpose of the canal was to afford an outlet for the products of the western part of the State to the east and to the lakes   In 1827 additional legislation provided for a portage road over the Allegheny Mountains, for the purpose of connecting the waterways of the two sections   From that time the internal improvements of Pennsylvania rapidly developed, and the bituminous coal trade assumed proportions of significant magnitude

In 1822, the Lehigh Canal, 108 miles in length, was constructed, and the development of the Lehigh coal fields was begun   The canal system was rapidly extended in various directions and Pennsylvania soon possessed what at that time were considered ample and adequate means of transportation

It was not, however, until the introduction of the steam railway that the coal fields of Pennsylvania became distinguished for their output   With the coming of the railways, they leaped at once into prominence and soon assumed that place of first importance to which they were justly entitled and which they are likely ever to maintain   The first railroad in Pennsylvania was a gravity road built in 1827, from Mauch Chunk, in Carbon county, to the Summit Mines, a distance of 9 miles   This was the first railway of any note constructed in the United States   The Philadelphia and Reading Railroad penetrated the same region in 1842, and was the chief factor in providing coal for the steam power that rapidly built up our industries and our commerce and advanced them to a position of acknowledged pre-eminence

The product of the bituminous fields about Pittsburg is distributed principally through its great waterway, the Ohio River, and the extensive systems of the Pennsylvania Railroad and the Baltimore and Ohio Railroad   The output from Bedford and Clearfield finds its way to market over the Pennsylvania and New York Central   The bituminous mining operations now extend over 25 counties in the western part of the State, and 33 different steam roads carry the product to market

The growth of the coal trade in Pennsylvania is best illustrated by figures   In 1820 the anthracite production amounted to 365 tons, in 1903 it amounted to 66,080,000 tons, representing a selling value at tide water of $307,360,000   In 1840 the bituminous product amounted to 464,826 tons, in 1903 it amounted to 105,000,000 tons, valued at $275,000,000

## COAL PRODUCTION

### Anthracite

| | | |
|---|---|---|
| 1820, | .. | 365 tons |
| 1840, | ..... | 1,064,914 " |
| 1860, | . | 10,488,168 " |
| 1880, . . . . | | 28,649,811 " |
| 1900, | . | 57,367,915 " |
| 1903, | . | 66,080,000 " |

Bituminous

| | |
|---|---:|
| 1840, | 464,826 tons |
| 1860, | 2,679,773 " |
| 1880, | 21,280,000 " |
| 1900, | 79,842,326 " |
| 1903, | 105,000,000 " |

Various estimates have been made of the amount of coal still remaining unmined in the anthracite fields  One computation places it at 10,638,902,809 tons, half of which may be available, or 5,319,451,404 tons  Another computation is 6,512,167,703 tons as available  Estimating the annual production at 60,000,000 tons, the duration of this industry in one case would be about 89 years, and in the other about 109 years

No estimates are obtainable regarding the bituminous deposits

The daily wages paid to miners in Pennsylvania have increased from 90 cents per day in 1840, to about $2 50 in 1903  In 1901 the anthracite miners produced an average of 464 short tons per man, the daily average production per man being 2 37 short tons  The bituminous miners produced an average of 664 short tons, the daily average per man being 2 94 short tons  The number of employes at the anthracite mines in 1903 was 151,479, and at the bituminous mines 160,000

The coal of Pennsylvania that has furnished the motive power for our vast commercial industries has made our State wealthy and populous  The number of our industries in 1900 was 52,185, and the value of the manufactured products was $1,835,104,400  Towns and cities, almost without number, have come into existence and grown with remarkable rapidity to a position of celebrity, since the general introduction of coal  The most prominent examples are the cities of Philadelphia, Pittsburg and Scranton  These cities contain, perhaps, more solid wealth in proportion to their population than any other cities in the country  Their prosperity, as well as that of many other Pennsylvania communities, is permanent because it is based largely on the creation of new values  Possessing unlimited quantities of coal as a creative power, they offer a most attractive field for investment and enterprise of every character

By the use of our coal we were enabled in 1902 to produce pig iron valued at $126,000,000, or almost 50 per cent of the total production of the United States, coke to the value of $56,700,000, rolled iron and steel to the value of $360 338,972

It is interesting to note that the value of the output of coal in Pennsylvania for 1903 exceeded greatly the total value of all the copper, gold and silver mined in the United States during that period  The copper output amounted to $88,334,770, gold $74,525,340, and silver $30,520,688, or a total of $193,380,798, while the total output of coal was valued at the mines at almost $300,000,000  At the seaboard the value was about $600,000,000

The control of our coal lands and various systems of railways, particularly in the anthracite region, is rapidly becoming centralized   In 1865 the anthracite coal lands were possessed by individuals, but in 1867 the railways began to acquire them and have steadily continued the acquisition until this vast industry is now under their domination   The further centralizing of these interests is being brought about by the merger of the different railroad companies under what is called "the community of interests" plan, and it will be but a very short time, if, indeed, it is not already an established fact, until one gigantic syndicate will control absolutely the destiny of the anthracite coal trade and with it indirectly the destiny of thousands of industries and countless thousands of human beings   The future welfare of the Commonwealth depends largely upon its coal trade, and the passing of the control of this source of creative power into the hands of a monopoly is a matter of great concern to the public in general   The forces, however, that compelled the consolidation in this instance were irresistible   Free and unrestricted competition had resulted in frequent disasters, as the past history of this industry plainly shows, and an agreement of some kind had become necessary.  The benefits to be derived from consolidation are, in the language of a recent writer on this subject   1  A better system of transportation, a more regular movement of the production, and a saving in the transportation equipment, which will reduce the cost of bringing coal to market   2  The doing away with coal agents and hence the cheaper marketing of coal   3  The working of the most profitable collieries and the shutting down of the more expensive ones   In this way, in mining coal under the most favorable conditions possible and using mining plants to the greatest profit, the cost of production will be reduced   4  Greatly reducing the cost of management by centralization, for much of the expense now incurred by divided interests in collieries and railroads would thus be avoided   5  The maintenance of prices at a steady scale which would yield reasonable returns to labor, management and capital

The dangers to be apprehended from this consolidation are restriction in production and raising of prices

The methods of mining in Pennsylvania differ very greatly in the b'-tuminous and anthracite fields   In the anthracite region three methods are now employed   One is stripping   This method, as its name implies, consists in stripping off the material overlying the coal and then mining it in the full glare of the sun   This is the cheapest method of producing coal

The second is the slope system   A slope is an incline plane driven from the surface down to or through the coal beds   It varies in dimensions   If it is intended for a single track, 12 feet wide by 7 feet high will suffice   If a double track is needed, then it must be about 22 feet by 7 feet

The third method is shaft mining   Seams lying 200 feet or more beneath the surface are generally worked by shafts, that vary in depth from 200 to 2,000 feet   In deciding the location of a shaft it is important to know the lay of the coal beds, and for this purpose bore holes are put down  Cross sections of the coal measures are thus secured, the geological structure ascertained, and the shaft located in the most advantageous place

In preparing anthracite coal for market, breakers are built at the mines for the purpose of cleaning the coal, removing the slate and other impurities, and of sorting it into sizes for the market  Breakers cost from $50,000 to $250,000 each

Other accompaniments of mining are the driving of headings to open up chambers, mules or other power to bring the coal to the foot of the shaft, carpenter and blacksmith shops, powder houses, and so forth

The mining of bituminous coal is much simpler  Little preparation is needed for the market, as it is sold without regard to the size and is often placed in the cars ready for shipment just as it comes from the mine  Bituminous coal is produced with much less effort than anthracite coal  The mines are usually entered from drifts, although there are some slopes, but they do not often extend below 300 feet

The difference in the expense of producing the two coals is very great In the anthracite region it is necessary to have a great army of slate pickers, tracklayers, pump runners, foot and slope men, drivers, doorboys, engineers, firemen, blacksmiths, carpenters, loaders and plate men, to mine the coal and prepare it for market  The effort, of course, with all  mine operators is to produce the greatest quantity of coal in the shortest space of time and with the least expense  Every conceivable device that has a tendency to this end has been adopted  The use of machines for the purpose of mining coal is rapidly coming into general practice  The modern mine owner, in his broad plan of conducting this industry, equips his mines with the finest mechanism and the latest improvements  We have now, in some of the mines, electric motors, coal washers, first motion engines, box car loaders, steel towers, detachable hooks for hoisting cables, and in the thicker seams machines for undercutting and drilling coal

A hundred years ago the miner, with pick and shovel and the aid of powder, labored long and hard to produce a ton of coal—twelve to fourteen hours  To-day, men work eight hours and with mining machines that undercut and shear the coal can produce from 3 to 10 tons The advent of machinery into the coal trade has greatly changed the conditions and has brought to that industry many men of high intelligence and broad mind, who realize the necessity of education among the class of people who operate the mines and who form the mining communities It is this that has led to the building of villages where the inhabitants can have comforts and some luxuries, and thus an encouragement to civic pride  School-houses have been erected and instructors provided, and under the laws of the State child labor is practically done away with and all children are compelled to attend school

The mining laws of Pennsylvania require the owner, operator or superintendent of every mine to provide and maintain sufficient ventilation to carry off and render harmless the noxious and dangerous gases generated in the mines  The ventilation in slope mines is effected by an inlet and an outlet  If the inlet comes down the slope a second opening is made at another point where the fan is attached, which creates a vacuum and so facilitates circulation  The law requires a fixed quantity of not less than one hundred and fifty cubic feet per minute for each person working in the lift  It is regulated by doors erected across gangways and other paths of air current, so that each working is provided with the necessary supply  In order to carry the air to the face of the chambers, new cross-cuts must be driven through the pillar at intervals of 30 feet  When a new one is driven, the old one is walled up tight

The expense of keeping up the ventilation in mines is considerable  In all the shaft mines the fans must be run night and day, regardless of the hoisting of coal  The airways must be carefully inspected every morning and evening  Hence there is a force of brattice men employed, whose duty it is to adjust doors, build partitions, wall up cross-cuts, and so forth, so that the visible air may be tractable to the needs of the colliery and led by divers ways to the workings  In the more gaseous mines the care bestowed on ventilation is ceaseless

Accidents in the mining of coal occur with appalling frequency, but the ratio to the number of employes in this industry is less than among railway employes  The unavoidable dangers are very great, but 50 per cent of the accidents are attributable to negligence and carelessness  The principal causes are falls of rock or coal, cars, powder and gas  In 1902, in the anthracite region, there were 300 fatal accidents and 641 non-fatal  The percentage of fatal accidents to each 1,000 employes was 2 700  In the bituminous region there were 456 fatal accidents and 861 non-fatal  The percentage of fatal accidents to each 1,000 employes was 3 368

Along with the development of the coal industry of Pennsylvania there has been ever-increasing attention given to the safety and welfare of the mine workers  Individual effort has been seconded by State legislation until to-day we possess not only the most modern machinery and efficient appliances, for producing coal, but we have surrounded this arduous and dangerous occupation with every safeguard and convenience that an intelligent liberality could suggest

The effort to raise the standard of intelligence among the mining communities and to afford opportunity for moral and mental improvement, has been continuous from the time the industry first assumed proportions of magnitude that made it a factor in our economic life  Higher wages, shorter hours, educational advantages, hospitals for the injured, and relief funds for the needy, have all been made the subject of thought and action by the individual and the State

This great industry, so vital to the welfare of the Commonwealth and so tremendous in its commercial import, is under the control and supervision of the State Department of Mines  The Chief of the Department receives his appointment from the Governor   James E  Roderick, of Hazleton, the present incumbent, is an intelligent, scientific miner   He has had practical experience in the mines, and for many years was a Mine Inspector in the anthracite region   He therefore possesses a thorough understanding of the duties and responsibilities of his position He has direct supervision of the thirty-one Mine Inspectors of the State, and it devolves upon him to see that in the discharge of their duties they satisfy all the requirements of the law

Pennsylvania still leads all other States in the magnitude of its production   America produces one-third of the entire coal tonnage of the world, and of this amount Pennsylvania has the distinction of producing about one-half   It is not a matter of surprise, therefore, that the keen-eyed capitalist and manufacturer has selected this State as the place of his industrial operations

In addition to the great fund of coal that will furnish the dynamic force for our mills and factories and numerous other industrial plants in the future, Pennsylvania possesses a system of railways perfect in organization and complete in equipment, offering to the manufacturer means of transportation at once rapid, safe and inexpensive

The wealth of a State consists first in its natural resources, and, second, in the intelligence and industry of its people   This axiom has in Pennsylvania a complete exemplification   Her resources consist of almost every species of raw material essential to make a State great in the three lines of development—agriculture, manufacture and commerce   By reason of these advantages, Pennsylvania has become a center of thought and intelligent progress, and offers to the capitalist not only the greatest opportunities for commercial investment, but the comforts and luxuries that are to be found only among a people of advanced ideas and exceptional culture

Printed in the USA
CPSIA information can be obtained
at www.ICGtesting.com
LVHW020754281223
767241LV00052B/331